ORILLIA CHRISTIAN SCHOOL
P.O. BOX 862
ORILLIA, ONTARIO

LEAPER

THE AMAZING LIFE of the SALMON

ULCO GLIMMERVEEN

Ashton Scholastic

Auckland   Sydney   New York   Toronto   London

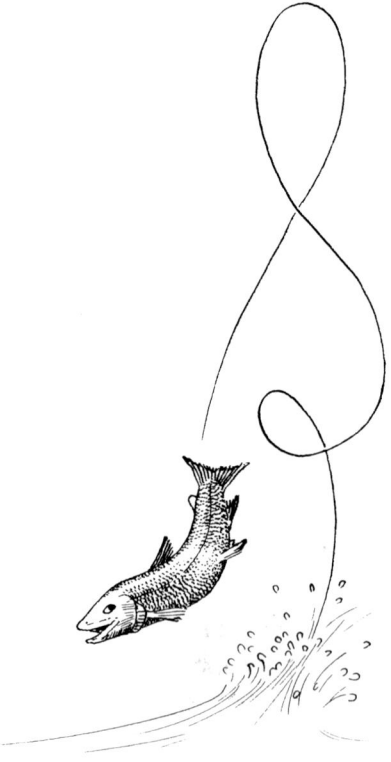

First published by Ashton Scholastic Ltd, 1991

Ashton Scholastic Ltd
Private Bag 1, Penrose, Auckland 5, New Zealand.

Ashton Scholastic Pty Ltd
PO Box 579, Gosford, NSW 2250, Australia.

Scholastic Inc.
730 Broadway, New York, NY 10003, USA.

Scholastic Canada Ltd
123 Newkirk Road, Richmond Hill, Ontario L4C 3G5, Canada.

Scholastic Publications Ltd
Marlborough House, Holly Walk, Leamington Spa, Warwickshire CV32 4LS, England.

Copyright © Ulco Glimmerveen, 1991

All rights reserved. No part of this publication may be reproduced or transmitted in any form or by any means, electronic or mechanical, including photocopying, recording, storage in any information retrieval system, or otherwise, without the prior written permission of the publisher.

National Library of New Zealand
Cataloguing-in-Publication data

Glimmerveen, Ulco.
   Leaper: the amazing life of the salmon / Ulco Glimmerveen. Auckland, N.Z. :
Ashton Scholastic, 1991.
    1 v.
   Picture book on salmon for children.
   ISBN 1-86943-042-5
   1. Salmon — Juvenile literature. I. Title.
    597.55

87654321   123456789/9

Edited by Penny Scown
Designed by Julie Roil
Typeset in Garamond Book by Rennies Illustrations Ltd

CONTENTS

INTRODUCING THE SALMON 5
    Origins 5
    Evolution 6
    Size 7
    What makes the salmon special? 8

THE SALMON'S LIFE CYCLE 10
    The egg 10
    The alevin 11
    The fry 12
    The parr 13
    The smolt 14
    The mature salmon 16
    Returning to the rivers 17
    Spawning 20
    Dying 22

WANDERERS 24

LANDLOCKED SALMON 25

THE SALMON'S SENSES 26
    Eyesight 26
    Taste 27
    Smell 28
    Hearing 30
    Pressure sense 30
    Temperature sense 31
    Sense of direction 32

CONSERVATION 33
    How the salmon became rare 33
    Saving the salmon 35

GLOSSARY 38

INDEX 39

BIBLIOGRAPHY 40

INTRODUCING THE SALMON

ORIGINS

More than a hundred million years ago the earth looked very different from the way it looks now. The continents were inhabited by strange plants and dinosaurs, and there were no people around. It was during this time that a group of fish called *Salmonids* originated.

EVOLUTION

At first, these primitive ancestors of the salmon lived only in what is now called the Atlantic Ocean, but after a couple of million years they found a passage through Arctic seas into the Pacific Ocean. They spread into this new area, but had to adapt to survive in such a different environment.

Each population of salmon gradually changed during millions of years of evolution. Finally, several groups (called species) developed, each slightly different from the others in size, colour, habits, and home range.

Nowadays, there are seven Pacific salmon species and one Atlantic species, all living in the Northern Hemisphere (although some species have also been introduced into Southern Hemisphere locations such as the South Island of New Zealand). The Atlantic species is simply called the Atlantic salmon. In the Pacific, however, are the chinook, coho, sockeye, pink, chum, cherry, and steelhead salmon.

SIZE

Usually, female salmon are larger than males, with the largest species being the Atlantic salmon and the chinook. When mature, they weigh around five kilograms on average, but can be much larger.
A huge chinook weighing 38 kilograms (as heavy as a large German shepherd dog) was once caught with a rod and reel. The largest salmon ever caught in a net weighed 51 kilograms, and was nearly 1.5 metres long.

This may explain why the chinook is sometimes called the king salmon. Such exceptionally large salmon are becoming increasingly scarce, however, with most weighing less than 20 kilograms.

The pink salmon is the smallest. It matures at two years old, at which time it weighs about 1.5 kilograms.

WHAT MAKES THE SALMON SPECIAL?

Although it is quite an ordinary looking fish, the salmon has many special qualities.

Its flavour

Many people are interested in salmon as a source of food. Each year, over 900,000 tonnes of salmon are caught in the oceans by commercial fishermen, and in the rivers they are an interesting challenge for sport fishermen. Salmon is considered a delicacy in many parts of the world, and is very expensive.

Its life cycle

The salmon's life cycle intrigues many people, including scientists. Salmon belong to a group of fish called *anadromous*, a Greek word meaning 'down-running'. After the eggs hatch in the swift-running streams high in the mountains, the young salmon 'run' down the rivers to the ocean. Years later they undertake a remarkable journey of thousands of kilometres to return and breed in the very same streams where they were hatched.

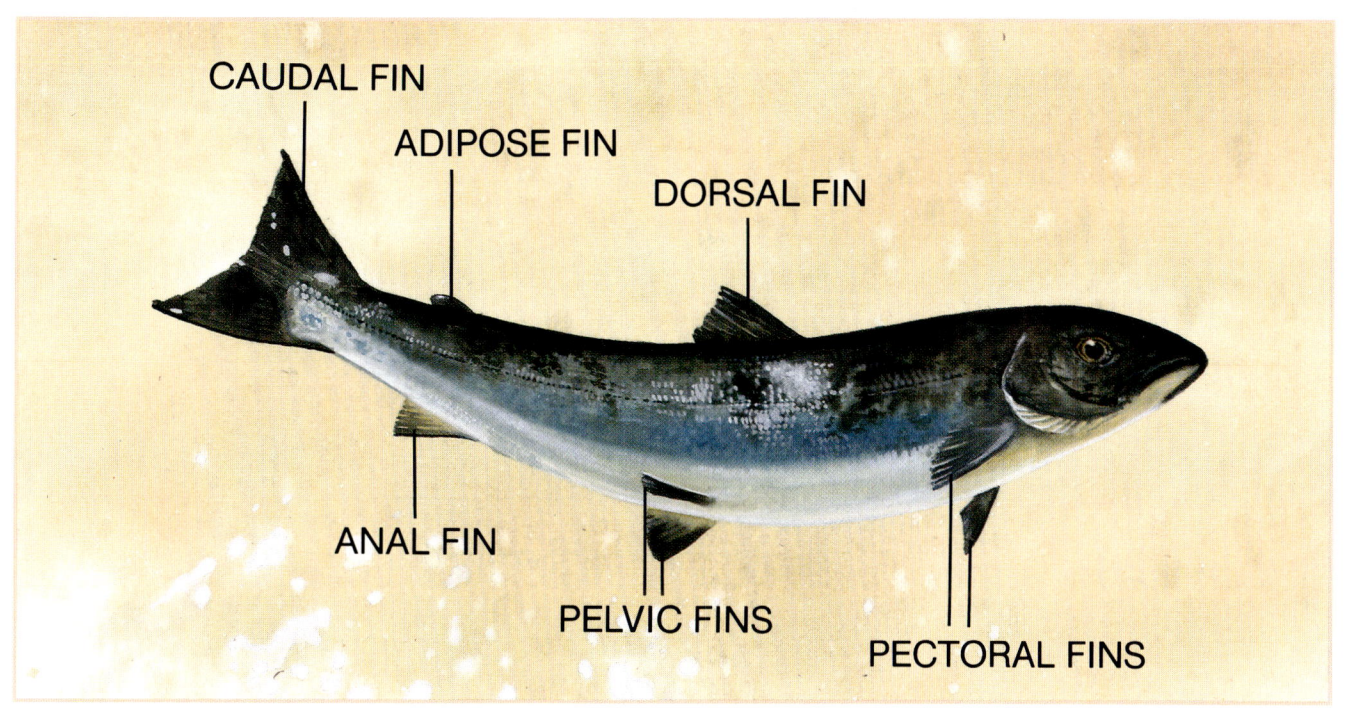

Its features and abilities

Salmon and trout are the only fish that have a small fin on their back, called the adipose fin. Apart from containing some fat, the adipose fin has no real purpose.

In total, salmon have eight fins. The median fins on the back and belly, called dorsal and anal fins, stop the salmon from rolling over, acting in a similar way to the keel of a sailboat. The two sets of paired fins, the pelvic and pectoral fins, prevent the salmon from falling forwards, and help it to steer. The tail, or caudal fin, is the largest and is used for movement.

A mature salmon has a long, slim body, and a small head. Its tail is very flexible and the tail muscles extremely powerful. This enables the salmon to travel distances of up to 20,000 kilometres in the ocean, and to reach speeds of 50 kilometres per hour. During its migrations upriver it can jump more than 4 metres (as high as a single-storeyed house) to ascend waterfalls. In America the salmon is also known as the 'leaper'.

The Salmon's Life Cycle

THE EGG

The life of a salmon begins in a freshwater stream where thousands of eggs, each the size of a pea, lie buried in the gravel. On average each female salmon lays 3000–7000 of these pink 'peas' in several nests in the streambed.

Many eggs die. They may be damaged, or unfertilized, or infected by fungi, and slowly they decay. At the end of the winter the rest of the eggs hatch.

THE ALEVIN

The tiny young salmon are called **alevins**. Each has a yolk sac attached to its belly which provides a perfect diet of proteins, sugars, vitamins and minerals for up to several months. The alevins stay safely in their nest in the gravel until this sac is used up.

THE FRY

When the yolk is completely absorbed, the alevin have become **fry** and emerge into the stream. They measure about 3–5 centimetres in length.

Hovering close to the safety of the streambed, the fry feed on microscopic organisms. Many are caught by waiting predators and do not survive. Large fish, (e.g. trout), waterbirds, insect larvae, and water beetles all feast on the thousands of fry.

The survivors gradually descend to larger streams and lakes. In most species of salmon the fry stay in the 'nursery lakes' for at least one year. During spring, food such as insects, larvae, worms and small crustaceans is plentiful, so the fry grow fast. They learn to avoid predators and to find the best feeding areas. The strongest fry may even defend a good territory against competitors. The weakest fry often gather together in schools.

THE PARR

Later during the same year the fry turn into **parr**, which is the next stage of the salmon's life. At this stage, parr are 5–10 centimetres long and have a brownish camouflage pattern. They are quite fierce, and are able to leap clear of the water to catch insects.

Parr of 10–20 centimetres long which have escaped the larger predators such as herons, otters and hunting fish, undergo a radical change the following spring.

THE SMOLT

The longer days of spring, the rising temperatures and the spring freshets influence the young salmon's bodies. Their sides turn silvery, and internally they adapt ready for a life in salt water. They become very restless. Once this process (called smoltification) is complete, the young salmon are called **smolts**. Of every hundred eggs only 1–3 survive to become smolts.

The smolts gather in schools and enter the rivers that run down to the estuaries. They are now ready to go to sea.

Once in the ocean, the smolts often travel thousands of kilometres, eating voraciously and growing rapidly. They usually stay in the upper 40 metres of the water, where they hunt for invertebrates and larval crabs and, later, squid, fish and crustaceans.

The salmon's dark back, silvery sides and white belly are perfect camouflage in the ocean. From below, the white belly and silvery sides are hardly noticeable against the sun's reflection on the waves above. And from above, the salmon's dark-bluish back is easily overlooked against the deep, dark water.

Life in the ocean is as dangerous as in the lakes. Many smolts are eaten by sharks, seals or sea lions. And, of course, many are caught in the huge driftnets of commercial fishermen.

THE MATURE SALMON

The smolts mature after 2–8 years of ocean life. At maturity, the salmon's instincts urge them to migrate back to the coastal waters.

Even after years of roaming the oceans, the salmon's extraordinary homing ability brings them back to exactly the same estuary they once left as smolts. There they assemble in schools and, while getting used to the fresh water again, their numbers begin to increase steadily.

RETURNING TO THE RIVERS

Soon the salmon venture into the rivers in great masses. Once in fresh water they stop eating altogether. While living on the fat reserves built up during their time in the ocean, their digestive organs shrink to make way for their fast-growing reproductive organs.

The salmon's colour changes from silver to a darker tone. They may even become brown, dirty yellow, greenish or bright red, depending on the species.

The migration upriver is very spectacular. Every year in Canada, thousands of people flock to watch the mature spawners struggle against the current. The salmon slither across rocks, run rapids and leap high into the air to overcome obstacles such as tree trunks, sandbanks and waterfalls.

The exhausting journey may use up almost all of the salmon's body fat and more than half of their muscle tissue. Many salmon are not strong enough to complete the trip, and die, or are killed by predators.

Although some species, such as the pink and chum salmon, travel only a few kilometres upstream, or even breed in the estuaries, the other species often run hundreds of kilometres upriver.

The longest spawning trip known is from the Bering Sea, up Alaska's Yukon River to Lake Teslin in Canada — an amazing distance of more than 3,400 kilometres. It takes these salmon two months to reach their parental streams.

SPAWNING

When the nights get frosty and the water cools in late autumn and winter, spawning takes place. By now, the remaining salmon are very colourful.

The males, called cocks, look particularly different. They have developed strange hooked lower jaws, called kypes, and long fierce-looking teeth. In some species they also develop a pronounced hump just behind their heads.

The stronger males have fierce fights to drive off the weaker males, while the females, called hens, fight for the best spawning places.

The hens roll their sides against the stream bed forming 10–30 centimetre-deep scoops in the gravel, while the male defends the female's territory.

When the female is ready, the male vibrates his body against her to stimulate her into laying the eggs. As soon as the female has laid her eggs, the male deposits his milt over them, and the hen fish moves a short distance upstream to prepare another nest, thereby covering the eggs in the previous one.

In total, a hen salmon lays approximately 1,700 eggs per kilogram of her body weight.

DYING

Most salmon use their last reserves of energy while spawning. A few days after the ritual is finished, the males and most of the females become listless. They drift helplessly with the current and quickly waste away.

Most Pacific species die sooner or later, but some strong steelhead and Atlantic salmon may survive to descend the river again and, after some years, return to spawn once more. Over a period of about ten years they may even return several times before eventually dying. The known record is six times (for an Atlantic salmon).

The dying salmon are washed up on the riverbanks. Sometimes the riverbanks in Canada are littered with dead salmon for kilometres, piled to heights of more than a metre. Many are eaten by scavengers foraging on the riverbanks. The rest decay and are eaten by microscopic organisms. Next spring, these organisms will be food for the new generation of salmon that emerge from the nests as fry.

WANDERERS

Most spawners return to their home stream because they recognise it from their youth. It is therefore the only good breeding place they know. But of every ten salmon, one or two end up breeding in another stream.

One explanation is that these wanderers may have been unable to recognise their home stream. Another possibility is that they stayed longer in the rivers while they were young, which gave them more opportunities to investigate other areas. These salmon know that other streams are also good for breeding, and that they don't necessarily have to return to the stream of their birth.

These wanderers are important for the species because they help to introduce new blood into other salmon populations. They may also find good, new breeding areas and establish new migration routes. If the old rivers are suddenly dammed or become too polluted, the wanderers' offspring in the new stream will still survive.

LANDLOCKED SALMON

Some populations of salmon breed in streams which flow into landlocked lakes, and their young never migrate to the sea. Instead, they descend to the depths of these cold lakes until they have matured, and then return to the streams to spawn. Usually, these so-called 'landlocked' salmon are much smaller than their oceanic relatives, probably because there is a greater abundance of food in the ocean.

Some salmon have been known to stay in lakes which are not landlocked rather than migrate down rivers to the sea — something which still puzzles scientists.

The Atlantic salmon has lake populations in the United States and Canada, where they are called quananiche, or sebago salmon, and this species has also been introduced into the southern lakes of New Zealand. The coho salmon has been introduced into the Great Lakes in the United States, and the kokanee salmon, a landlocked form of the sockeye salmon, lives in several lakes in the United States.

THE SALMON'S SENSES

The life of a salmon is not easy. Of every 4,000 eggs, only two live to be mature spawners.

How do salmon survive the hazards of their journey? How does a salmon avoid predators, and, most importantly, how can it ever find its way back to its home stream?

Salmon have developed some extremely perceptive senses, which are the key to their survival. But a lot of the salmon's extraordinary orientation ability is still a mystery. Let's look at what scientists have discovered to date.

EYESIGHT

The salmon's eyesight is very important for hunting and for enabling them to assemble in schools. It also helps them to recognise landmarks in rivers and lakes and in their home streams. To help orient themselves in the ocean, they are able to use the light of the sun and moon as beacons.

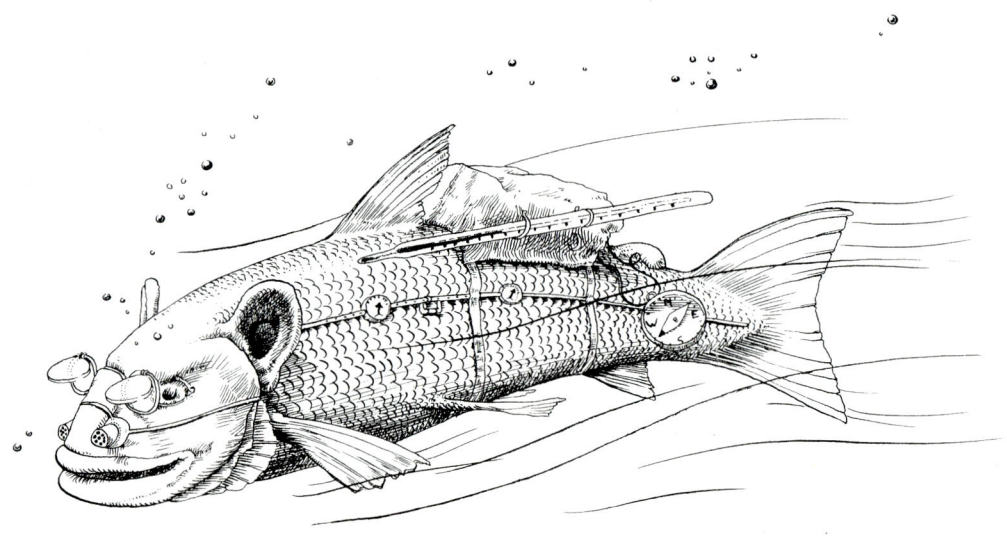

26

The salmon's sharp vision is limited to about 30 metres, but this is all they need underwater. They see their surroundings in colour. They can even increase their sensitivity to the bluish light of the ocean once they have migrated from the shallow rivers into the deep sea. This enables them to see the best they can in the prevailing light.

TASTE

Like humans, fish are able to taste sweet, bitter, sour, and salt, but in many fish species the sensitivity is much greater — up to a hundred times greater, in fact. Their taste cells are found not only inside the mouth, but also on the lips and snout.

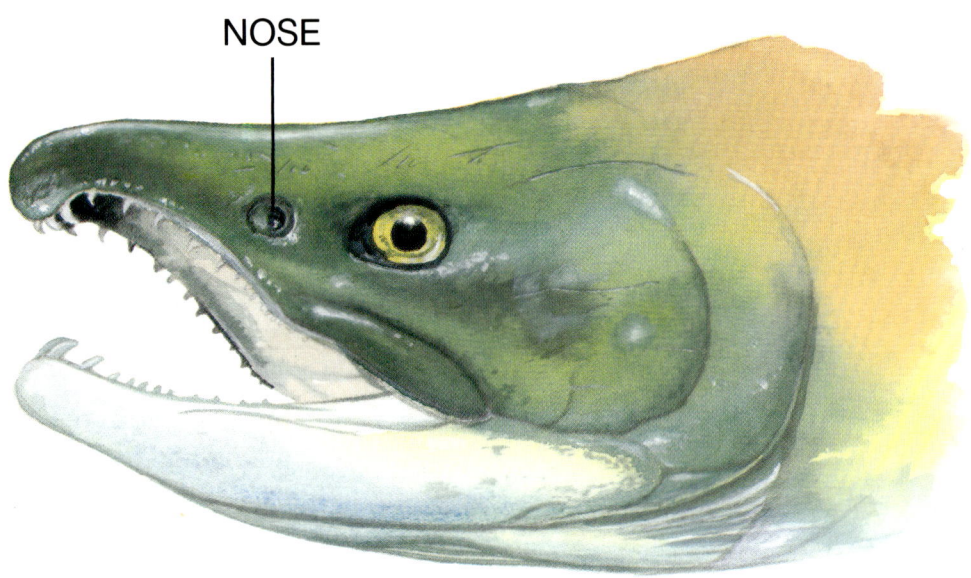

SMELL

A sense of smell is particularly important to migrating fish such as salmon. The nostrils of a salmon's nose are tubes connected to a chamber which is filled with sensory cells. These sensory cells are extremely sensitive, even more so than those of a highly trained tracking dog, and can readily distinguish between different smells in the water. Salmon probably smell with other parts of their body too, such as in the skin of the gill covers, belly and tail.

Salmon are far more dependent on their sense of smell than we could ever imagine. The vegetation of the surrounding area gives every stream a particular scent. A young salmon becomes familiar with this.

Years later, when the salmon migrates back to the coast as a mature spawner, it recognises this scent where its home river flows into the sea. From there, the salmon simply follows the trail of scent through the estuary into the river, through the lakes and eventually into the stream where it was born.

At spawning time, salmon can identify males from females by their scent. Predators such as bears, dogs, seals and humans each emit a particular scent. If any of these salmon hunters are waiting somewhere in the shallows, the approaching salmon will recognise their scent and may take a detour to avoid capture.

Another very interesting feature evolved in the ancestors of the salmon about 70 million years ago. If a salmon is injured by a predator, its skin releases a 'shock substance'. Other salmon that pick up this scent are thereby warned of danger and may be able to escape the enemy.

HEARING

Salmon's ears are not exposed as our ears are, but are hidden deep inside the salmon's head. Even so, their ears can still pick up underwater sounds, such as the voices of whales and dolphins, waves breaking on the rocks, or the throb of a boat's engine.

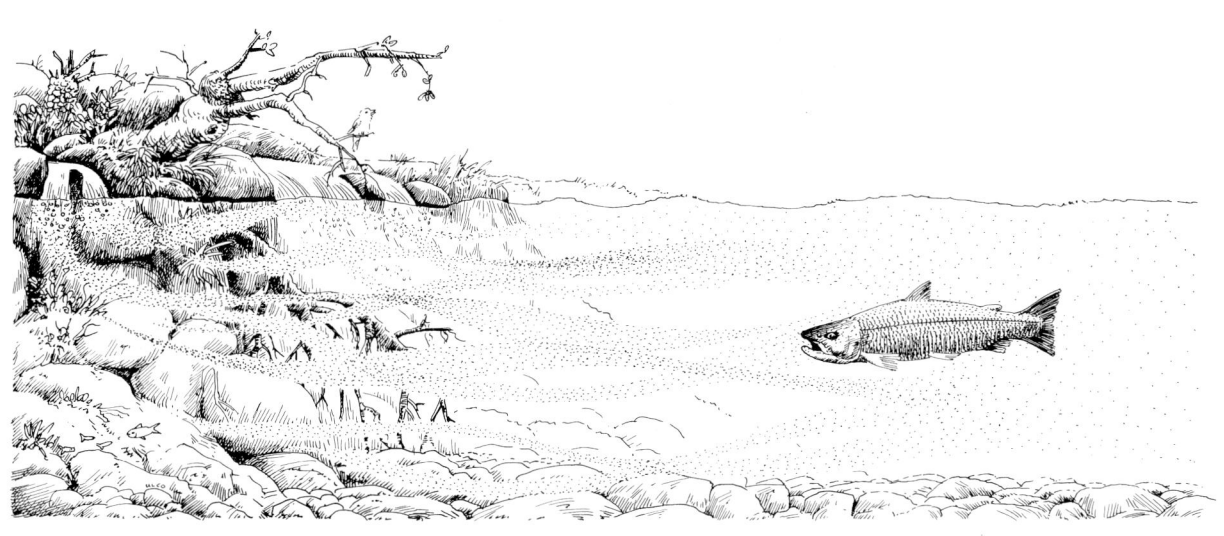

PRESSURE SENSE

Like most other fish, salmon have marks on the sides of their bodies. Some people fond of eating fish believe that these lines are there to show where best to cut it up! In actual fact, they are very sensitive 'lateral-line' organs, through which the fish are able to feel vibrations, or pressure waves, in the water.

There are always many vibrations in the water, caused by movements of other fish, prey, predators, ships and currents. A salmon can *feel* a predator's approach, and it can *feel* the direction of an ocean current.

Scientists think that salmon are able to use these currents for orientation. In the ocean, currents occur in quite regular patterns. After a few years, the salmon may come to recognise the pattern and know where the currents are and where they are going. Often, the salmon probably follow currents to make their journey less tiring.

TEMPERATURE SENSE

As with many other fish, salmon are very sensitive to temperature changes. If their heads touch water that is as little as 0.03° Celsius warmer than the rest of the water, they will notice it immediately. Salmon can even learn to recognise a certain temperature with complete accuracy to within one degree. When the temperature of their nursery streams rises in spring, the fry notice it and are stimulated to migrate down to the lakes.

In the ocean, being able to distinguish currents of different temperatures can be useful to the salmon. A certain current may lead them to rich feeding areas, or back to the coast.

SENSE OF DIRECTION

Without using any clues such as the sun and moon, currents, or scent, salmon still have a remarkable sense of direction.

They appear to contain an organ which could be compared to a built-in compass. This special organ can detect the earth's magnetic field, enabling the salmon to distinguish north from south, and east from west. Recently, scientists have found that salmon contain magnetite, a naturally occurring mineral, which could be the basis of this compass.

CONSERVATION

HOW THE SALMON BECAME RARE

Two hundred years ago, hundreds of rivers in Asia, North America and Europe were highways of salmon migration. Hundreds of millions of mature salmon returned from the ocean and travelled these rivers each year. Since then, however, much has changed.

Pollution of the rivers had already begun 150 years ago as more and more factories were built, discharging waste into the rivers. The chemicals released into the water masked the scent of the salmon's home streams, and the salmon were unable to find their way back. As the pollution became worse, breeding and nursery waters became contaminated. Chemical pollutants also decreased the oxygen content of the water, suffocating the fish. At sea and in the rivers, oil spills killed many more salmon.

Other problems encountered by the returning spawners were destroyed wetlands, waterside logging and engineering works. Dams were built and rivers diverted, causing the rivers to swell or thin. This often made it difficult, or even impossible, for the salmon to reach their natal spawning grounds.

And since commercial fishermen discovered the salmon's feeding areas in the ocean, salmon have been caught in massive numbers in driftnets of up to 50 kilometres long. In addition, poachers caught many of the returning spawners in large nets in the estuaries and rivers. The number of salmon being caught worldwide soon declined.

Nowadays, salmon runs are becoming a rarity, especially in Europe.

SAVING THE SALMON

Fortunately, measures are being taken to save the salmon. Commercial fishing is controlled: fishermen may catch only a set number of salmon each year. New laws are being introduced to ban the driftnets, also called 'walls of death', because they kill everything in the ocean which becomes entangled in them.

Cleaning of the rivers has proved to be very successful. About thirty years ago, the River Thames which runs through London, England, was completely devoid of fish. In 1959, work began in an effort to cleanse the river of pollution. Factories were forced to clean their waste, and sewage had to be purified before it was released into the river.

Gradually, the water quality improved and many species of fish returned. In 1974, sixty years after the last salmon had been seen in the dirty Thames, the first salmon of a new generation was caught. Since then, more salmon have returned to spawn.

Salmon are also being reared in hatcheries. They are caught in the estuaries and their eggs are hatched and kept safe until the young can survive. New populations can then be reintroduced into rivers from which the salmon have disappeared. One of these rivers is the Rhine in Europe, which was one of the best runs in the world until 1900. Current studies being sponsored by a chemical factory hope to predict the possibility of salmon being back in the Rhine by the year 2000.

Dammed rivers can still be made suitable as salmon runs by building a 'salmon ladder' — a series of low dams incorporated into the major dam, like a staircase. The salmon can leap from one step to the next and continue their journey upriver.

Salmon watching is becoming a popular tourist attraction. In rivers with healthy runs, many thousands of people come to watch the leapers run and breed. This increasing public interest may help to enforce strict control on commercial fishing, the cleansing of rivers, and the restoration of salmon runs.

If we can ensure the return of this vulnerable species into once-polluted waters, the salmon will stand as a symbol of hope for the future preservation of our natural heritage.

GLOSSARY

catchment	the area around a stream in which rainwater collects
crustaceans	a group of hard-shelled animals, e.g. shrimps, crabs, and lobsters
driftnet	a fine-mesh fishing net, weighted at the bottom and supported with floats at the top
estuary	the wide mouth of a river
freshet	a sudden rise in the level of a stream, or a flood, due to heavy rains or the rapid melting of snow
fungi	a special group of plants, e.g. moulds, mildews, and mushrooms
home range	the area in which an animal or species lives
homing ability	the ability to find the way home
invertebrates	a group of animals which do not have a backbone, e.g. insects, worms, and snails
landmark	a prominent feature of the landscape (including underwater) marking a specific place
median	the mid-point dividing two equal halves
milt	the reproductive sperm of male fish
nursery	the place where young animals or plants grow up
parental stream	the stream where a salmon's parents bred
population	all the animals of a species which live in a certain area
predator	an animal that feeds upon other animals
scavenger	an animal that eats refuse or decaying matter
sewage	the waste matter which passes through sewers
spawning	producing and depositing eggs and sperm
territory	the area occupied by an animal or pair of animals, usually protected and enclosing the most favourable area of food availability
wetlands	an area in which the soil is saturated with, or under, water — such as swamps, marshes and lake fringes

INDEX

alevin, 11
appearance
  of parr, 13
  of salmon, 14, 17, 20, 30
  of smolt, 14
Asia, 33
Atlantic Ocean, 6
Atlantic salmon, 7, 22, 25
breeding, 10–14, 20–22, 24, 25
Canada, 18, 19, 23, 25
cherry salmon, 7
chinook salmon, 7
chum salmon, 7
coho salmon, 7, 25
conservation, 33–35
death, 10, 22–23
driftnets, 14, 34, 35
eggs, 10, 14, 21, 26
England, 35
Europe, 33, 34, 35
evolution, 6
females, 7, 10, 20–21, 29
fins, 9
fishermen, 8, 14, 34, 35
food
  of alevins, 11
  of fry, 12, 23
  of mature salmon, 17, 25
  of smolt, 14
  salmon as, 8
fry, 12
kokanee salmon, 25
lakes, 12, 25, 28
landlocked, 25
life cycle, 8, 10–16
males, 7, 20–22, 29
migration, 8, 9, 16–19, 26, 27, 28, 30, 31, 32, 33

milt, 21
nests, 10, 11, 21
New Zealand, 7, 25
Pacific Ocean, 6
parr, 13
pink salmon, 7
pollution, 33–34, 35
predators, 12, 13, 14, 18, 29, 30
rivers, 14, 17–19, 23, 24, 28, 33, 34, 35, 37
runs, 8, 34, 37
scoops, 21
sebago salmon, 25
senses
  direction, 16, 30, 31, 32
  eyesight, 26–27
  hearing, 29
  pressure, 30
  smell, 28–29, 33
  taste, 27
  temperature, 31
size, 7
  of fry, 12
  of parr, 13
  of mature salmon, 7, 25
smolt, 14
sockeye salmon, 7, 25
spawning, 19, 20–22, 25, 29
species, 6–7
speed, 9
steelhead salmon, 7, 22
streams, 8, 10, 12, 19, 24, 25, 26, 28, 31
survival, 12, 26, 35–37
tail, 9, 28
wanderers, 24
weight, 7

BIBLIOGRAPHY

Attenborough, David *The Living Planet.* Collins BBC, London, 1984.
Caras, Roger *The Endless Migrations.* Truman Tally, New York, 1985.
Caras, Roger *Sockeye.* The Dial Press, New York, 1975.
Cloudsley-Thompson, John *Animal Migration.* Orbis Publishing, 1978.
Credland, Peter *The Living Earth: Rivers and Lakes.* The Danbury Press, 1975.
Droscher, Vitus B. *The Magic of Senses.* W.H. Allen, London, 1969.
Durrell, Lee *Natuurbescherming in aktie.* M & P Weert, 1987.
Dyk, Jere van 'Long Journey of the Pacific Salmon'. *National Geographic Magazine,* 1990.
Gibbons, Boyd 'The Intimate Sense of Smell'. *National Geographic Magazine,* 1986.
Harden Jones, F.R. *Fish Migration.* Arnold, 1968.
Hardy, A.C. *The Open Sea (2).*
Hvass, Hans *Fishes of the World.* Methuen, London, 1965.
Lee, Art 'The Leaper Struggles to Survive'. *National Geographic Magazine,* 1981.
McKeown, Brian A. *Fish Migration.* Croom Helm/Timber Press, 1984.
Marshall, N.B. *The Life of Fishes.*
Migdalsky, C. & George S. Fichter *The Fresh and Saltwater Fishes of the World.* Vineyard Books, New York.
Wheeler, Alwayne *Fishes of the World: Illustrated Dictionary.* MacMillan Publishing.
Co-authored *Fishes.* TimeLife, 1978.
Co-authored *Nightwatch.* Roxby Nightwatch Ltd, 1983.